PORTABLE COMPOSTING MACHINE WITH MECHANICAL\ELECTRIC SHREDDER

DR. VIJAY JHAMB & ER KIRANPREET SINGH

TABLE OF CONTENTS

CHAPTER 1 4
  INTRODUCTION 4
    1.1 RESEARCH BACKGROUND 4
    1.2 PROBLEM STATEMENT 5
    1.3 OBJECTIVES 5
    1.4 SCOPE OF RESEARCH 6
    1.5 SIGNIFICANCE OF RESEARCH 6
CHAPTER 2 7
  LITERATURE REVIEW 7
    2.1 PROJECT BACKGROUND 7
      2.2 PREVIOUS WORK 10
      2.2.1 System name: - WEMI-4000 10
      2.2.3 Ridan Composter 12
      2.3 COMPARITIVE STUDY 13
CHAPTER 3 15
  RESEARCH METHODOLOGY 15
    3.1 INTRODUCTION 15
    3.2 RESEARCH DESIGN 16
    3.3 MATERIAL SELECTION 17
    3.4 FABRICATION 19
    3.5 POWER TRANSMISSION PART 21
CHAPTER 4 24

FABRICATION PROCESSES..24
4.1 PROCESSES INVOLVED: 24
    1.Cutting: 24
    2.Drilling: 25
    3.Welding 25
    4.Grinding:.. 26
    5.Bending: 27
4.2 ENGINEERING DESIGN STANDARDS 28
    4.2.1 Screws and nuts 28
    4.2.2 Sprockets and shaft 29
    4.2.3 Motor 30
    4.2.4 Temperature Sensor 30
4.3 THEORITICAL CALCULATIONS 32
4.4 RESULTS 32
CHAPTER 5 33
5.1 LIFE LONG LEARNING 33
    5.1.1 Hardware Skills 33
    5.1.2 Time Management Skills 34
    5.1.3 Project Management 34
5.2 IMPACT OF ENGINEERING SOLUTIONS 34
    5.2.1 Society 34
    5.2.2 Economy 35
5.2.3 Environment 35
5.3 Contemporary Issue 35

CHAPTER 6 36

DISCUSSION AND CONCLUSION...36

6.1 CONCLUSION 36

6.2 FUTURE RECOMENDATION 36

REFERENCES 36

CHAPTER 1

INTRODUCTION

1.1 RESEARCH BACKGROUND

India is the second largest producer of fruits and vegetables in the world (after China) with 241.43 million metric tons. In a country like India, waste management and disposal of the waste to be a tedious task where a huge number of organic wastes has been generated due to enormous consumption. There also lies an issue in transporting the wastes to the recycling plants, wherein a huge amount of effort, time and money needs to be put in for transporting the wastes which are clustered and are not in uniform shape or size. Large amount of organic waste is produced in houses which eventually goes to the landfill and produces methane a major global warming & climate change gas. There lies an opportunity of converting the household organic waste into useful compost using portable composting machine.

Decomposition of organic waste in landfills is of anaerobic type and is not climate friendly as it produces methane which is a potent global warming gas. This problem can be solved itself in house by aerobically composting the microorganism convert shredded organic waste into compost in a temperature-controlled environment.

The shredder would shred the organic waste thus increasing the surface area. Higher the surface area higher is the rate of decomposition.

The shredder is designed to operate by both mechanical and electrical means. This shredder is designed in such a way that it is simple to construct and would require minimum effort for operating in both mechanical and electrical types of operation.

1.2 PROBLEM STATEMENT

Composting process needs finely shredded organic waste and temperature control environment. Problem is that in natural aerobic composting temperature may suitable for crore households, so average of 22.5 crore kg or 2.25 lakh tones of organic waste per day is produced. When household organic waste goes to landfill it undergoes anaerobic fermentation so methane is produced which is a potent global warming creating gas. We all know that the waste of food unavoidable, so the best way to makes this food waste useful is by composting the organic waste and returning to the environment. There is a need of such device which shreds organic waste and also provides suitable temperature for composting.ie below 35 degrees Celsius.

1.3 OBJECTIVES

1 Eco-friendly mechanical device which turns household organic waste to compost and stops the organic waste from getting to landfill

2 Providing temperature- controlled environment so that aerobic composting process takes place properly. This is achieved using temperature sensors and cooling unit.

3 Green Innovation, the need of the hour to fight climate change.

4 Stopping organic waste from getting to landfills and making value added product from it. Compost Produced can be used for farming.

5 Waste management of household waste.

1.4 SCOPE OF RESEARCH

1. It can be used in household setups for composting organic waste and its management.

2 Size of compost bin can be varied with increase in amount waste keeping same shredder.

3 It can be used in buildings and societies for waste management and compost produced.

1.5 SIGNIFICANCE OF RESEARCH

With such a huge amount of waste generated every single day both at a domestic level and industrial level, it becomes vital to handle waste effectively. The food dumped into landfills release **methane (CH_4)** which is dangerous for the earth as a whole. Composting is the best way to handle the waste. Instead of dumping more and more waste into the backyard and landfills, it is wise enough to turn waste into usable manure/compost. With technological advancements, it has been possible to even accelerate natural processes. Industries and businesses are shifting towards automation. An organic waste composting machine is an independent unit that facilitates the composting process and provides better composts. It takes waste as its input and provides manure as its output. The composting machine **accelerates the composting.** Shredding materials increases the surface area on which microorganisms can feed. Smaller particles also produce a more homogeneous compost mixture and temperature control unit to help maintain optimum temperatures.

CHAPTER 2
LITERATURE REVIEW

2.1 Project background

Composting is known as a natural process, it occurs by using microorganism under specific condition, which leads to the decomposition of organic waste. As we know, food waste one of the biggest global issues that faces the world nowadays, it could be at home, school, restaurants and any food service sector. One of the recent statistics said that billion tons of food waste is generated every year. We all know that the waste of food unavoidable, so the best way to makes this food waste useful is by composting the organic waste and returning the nutrients back into the soil to make the cycle of life to continuous which will help protecting of environment.

Figure 2.1 Food waste

There are two common types of composting process that can be used, anaerobic and aerobic. Aerobic composting produces methane (the main by-product of anaerobic degradation). The heat is produced in aerobic composting is sufficient to kill microorganism as these organisms are not adapted to these environmental conditions so, composting machine needs a temperature control unit. This composting process takes 8-10 days.

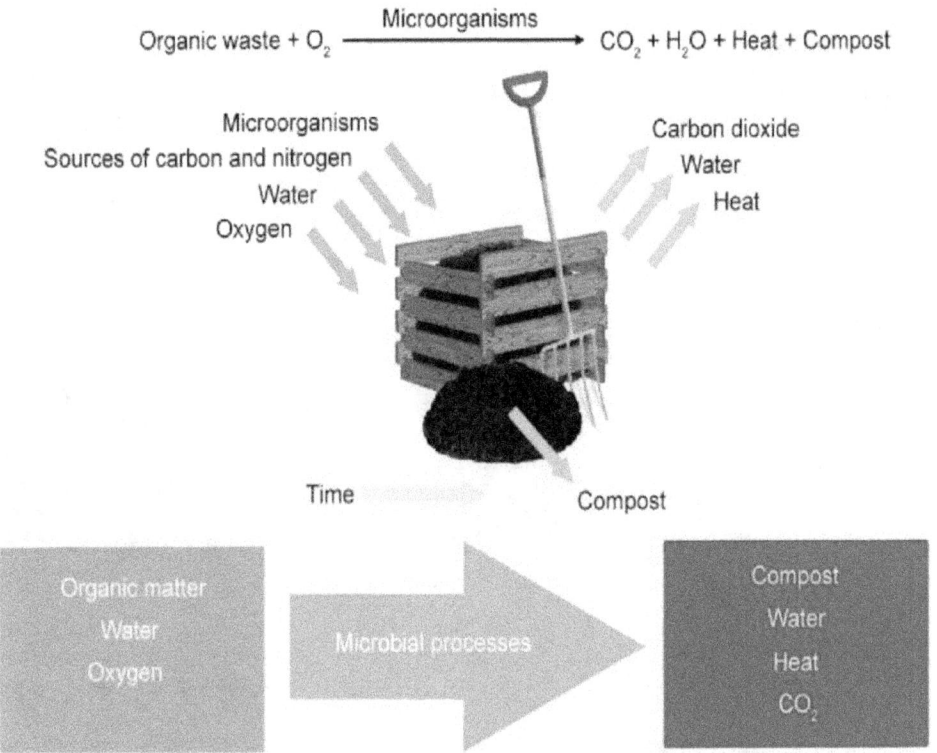

FIG 2.2 Composting process

The second type called "Anaerobic" and this composting process use no oxygen. The composting occurs by making the organic waste piled up and starts breaking down by its self. This process is very slowly and takes many years in order to compost the organic waste. The process is characterized by very strong odors.

Compost is a valuable soil amendment, as the compost feeds soil. The use of compost is an effective way to improve plant growth. Compost can be used for bioremediation of soil and pollution prevention, reduce erosion and nutrient runoff, alleviate soil compaction.

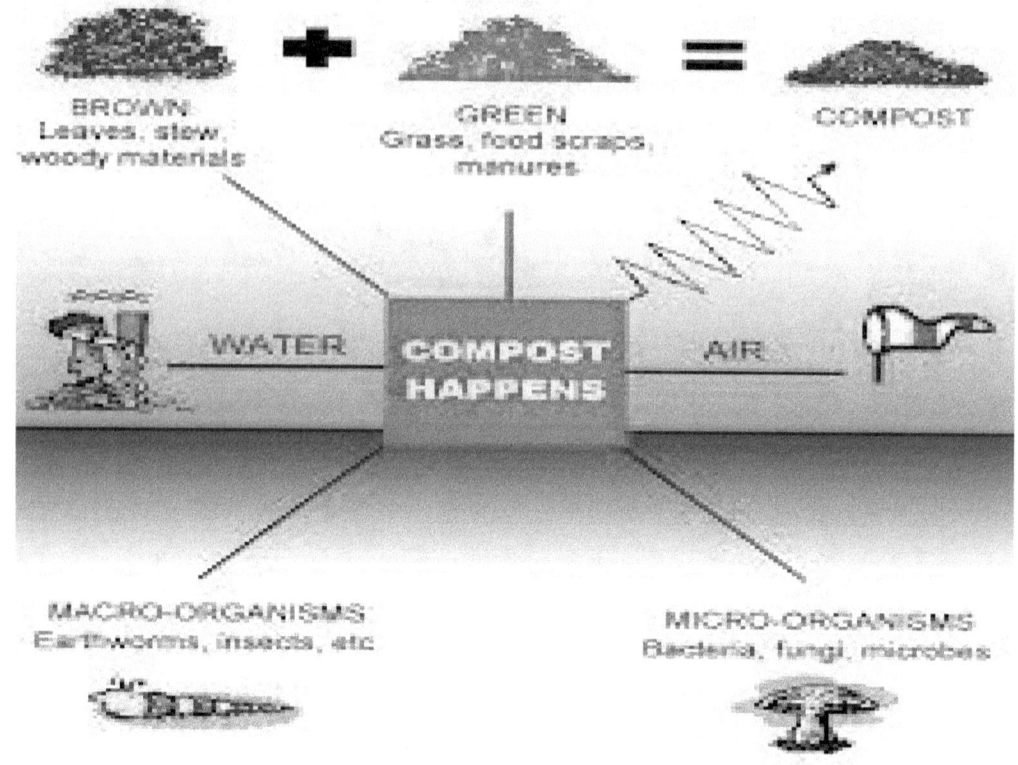

FIG 2.3 Aerobic Composting

Composting also helps the soil retain much needed moisture, and research has also shown that composting can also assist in enhancing the disease resistance of some plants, like tomatoes and vegetables. This can reduce the number of crops you lose to disease, which often leads to wasted expenses.

Aerobic Composting is most suitable for composting in household system because of fast composting.

Aerobic composting takes place in temperature below 35 degrees.

Aerobic composting takes time of 14 days for composting.

2.2 PREVIOUS WORK

In this section we tried to look for different composting methods and machines in order to find the capability and how they work, these subsections below are examples of different composting process that we will discuss in details in order to be able to compare them to our designed machine.

2.2.1 System name: - WEMI-4000:

Ohio University became the university with the largest in-vessel compost facility in the United States with the (Wright Environmental Management, Inc WEMI-1000) that was installed in 2009 a 2 ton in-vessel composting system. The tunnel inside the system is controlled for air supply and temperature a supply and exhaust fan and an air circulation Figure (2 2) shows that Composting material moves through a set of spinners that act to invert homogenize agitate and stack the material into the next zone.

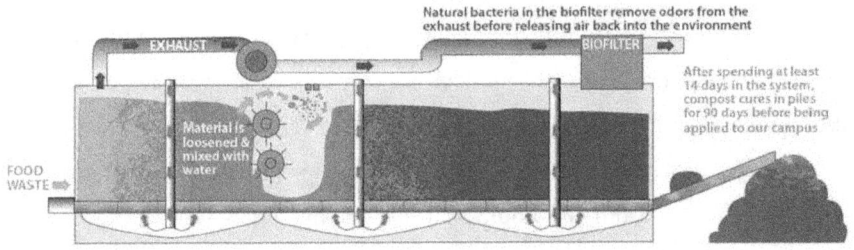

FIG 2.4 Ohio in vessel composter

Water will be added to the mix during material cross-mixing (if needed) to raise moisture levels into the desired levels Maternal remains in the second zone for an additional number of days equivalent to the retention time m Zone 1.

Flows into the pipes that run along the base of the tunnel and from the plenums to sump boxes through pipes located at the sides of the tunnel. Leachate is pumped back onto the composting materials from the sump boxes through pipes.

2.2.2 The Earth Flow:

The St Peters university has invested in a fully automated composting system called the Earth Flow This composting machine is located at the Foothills campus Finished compost is used in landscaping projects on campus.

The Earth Flow Capacity is up to 900 Kg of waste per day. That Waste is loaded into one end of the vessel by placing the collection container on an automated tipper Every time food waste is added, bulking material is added in a 1.2 ratio. Straw, wood chips and horse manure from the Foothills campus are the primary bulking materials. The inclined auger mixes and advances the compost down the vessel with each pass. The control panel allows the operator to select the number of times per day that the compost is mixed as well as automatically adds moisture to the compost.

Material composts in 14-21 days inside the machine. The auger discharges the finished compost through an end door of the vessel. The compost is to cure for at least 3-4 week. Figure shows that heat is created when micro-organisms, (including bacteria and fungi) break down the organic matter, bio food waste and wood. The heat attracts even more hyper active microbes, which make the composting process quick and efficient. This creates the perfect composting conditions. The food waste needs to stay inside the composter for a minimum of two weeks before it can be removed from the composter Depending upon what sort of food is being composted it may be ready to spread straight onto the garden. However, it is usually best to mature the compost for 23 months in a maturation Box.

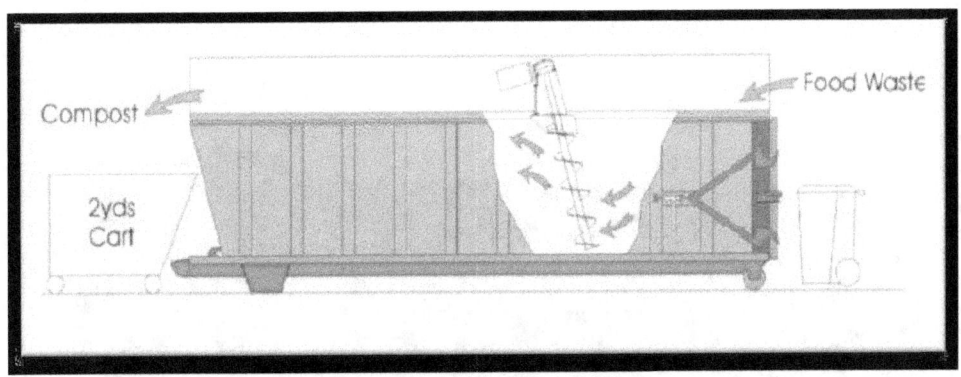

FIG 2.5 Earth flow composter

2.2.3 Ridan Composter

The Ridan food composter uses natural ingredients and processes to create a warm environment in which food waste (nitrogen) and wood (carbon), can mix with air and water Unlike other composters. all this happens without the need for electricity. making your Ridan cheap and easy to use.

FIG 2.6 Ridan Composter

2.3 COMPARITIVE STUDY

We have identified many factors that should be looked at when comparing composting systems and machines but we will focus on what makes our designed project considered superior to the other two projects that we have discussed in section 2.2.

One of the most important parameters in the composting process is the process time, if we took a closer look we can see that the (Wright Environmental Management, Inc. WEMI-4000) system that was installed in Ohio university can process organic waste in no less than 14 days, after that the compost shall cure for at least 90 days before it can be used as soil fertilizer, while the Earth Flow composting system that is being used in Colorado university can compost the organic waste in 14 to 21 days.

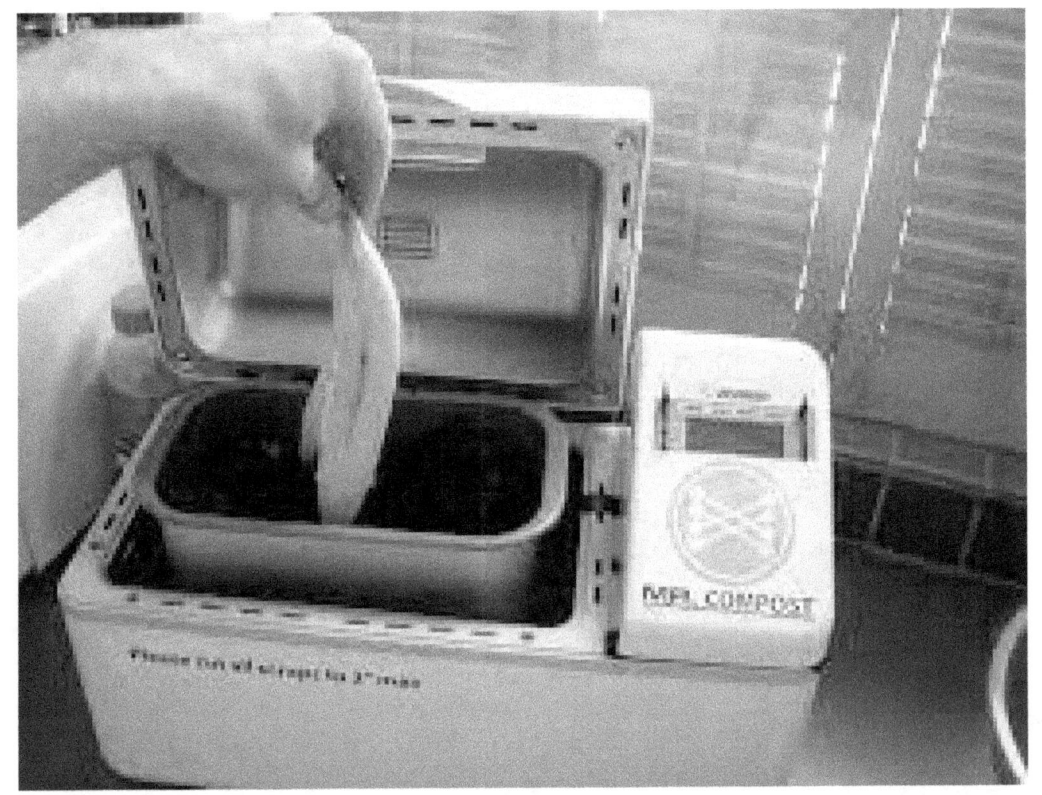

Fig 2.7 Digital Composter

CHAPTER 3

RESEACRH METHODOLOGY

3.1 Introduction

This machine can extensively used in homes, schools and societies. Beneficiaries include anyone who produces organic waste. This machine has been reduced to smaller scaleto get the best results while composting. It can be used in houses for making compost, usually people throw away their organic waste in garbage which goes to the landfill and produces methane. This machine will provide in house solution of organic waste management and in return produce compost for garden. Societies can set up these machines on every floor.

3.2 Research design

We discussed about the design, for a considerable time and finally the best design has been made. This design has been made to provide in house solution of composting organic waste to people who live in metros and do not have access to land for pit composting. The design is based on feedback from people who compost organic waste and solution of problem they face is incorporated in this project. Machine is made portable and user friendly and can be operated by person age more than 14 years. The machine also provided assurance of organic waste management + compost production.

3.3 Material selection

1. Plywood:

FIG 3.1 Plywood

Used in making frame of shredder

2. Alloy Steel:

FIG 3.2 Sprocket

Sprocket, used a blade in the shredder

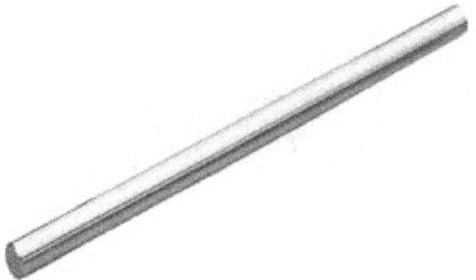

FIG 3.3

Shaft, used in power transmission from motor

3. Stainless steel:

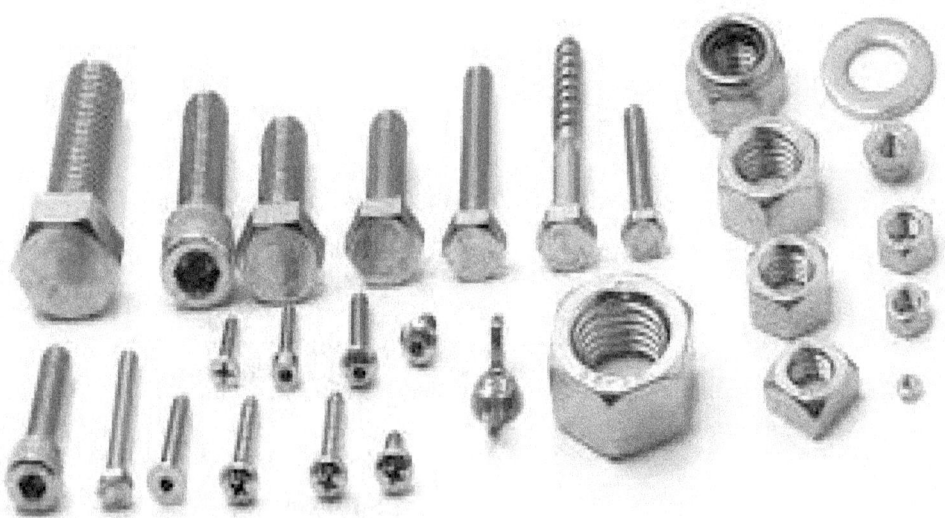

FIG 3.4 Nut Bolt

Nut bolts, washers for holding parts together rivets for making space in sprockets joined together.

4. PLASTC, RUBBER, COPPER

FIG 3.5 WIRES

Used in rubber connecting wires with copper core, POLY VINYL CHLORIDE (PVC) PCB sensors, Polyuthrane (PLASTIC) composting bin,

3.4 Fabrication:

In this project, a lot of fabrication work need to be applied to make sure the project done well and satisfying. The main adhesive agents that have been used are welding, which ARC is welding. The main reason arc welding is applied to is applied to the project is because it is an easy welding process compared toothers. Other than that, arc welding also produces a good weld surface and clean.

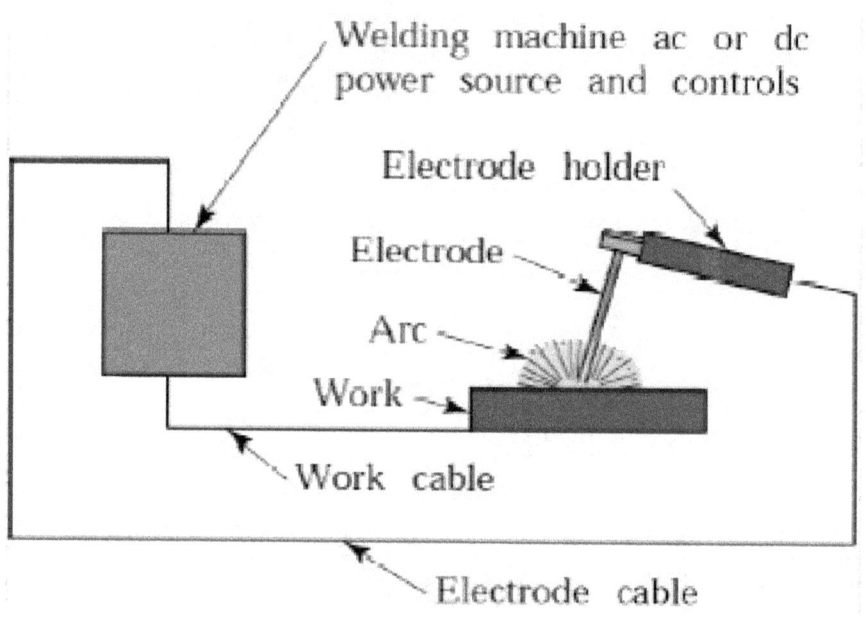

FIG 3.6 Arc Welding

Flow chart for Mechanical shredder machine

The figure shows the assembly procedure of coconut leaves shredder. Quality of the material has been checked at purchase level to meet the design needs. In this step a skeleton of the section is fabricated according to dimensions mentioned as per the design Fig 1: Flow Chart for Agricultural waste shredder machine In this step a skeleton of the section is fabricated according to dimensions mentioned as per the design. Frame is fabricated according to specified design and material. Then move on to cutter assembly here first to prepare the cutter container after that inset the shaft then cutter is mounted on shaft with key and spacer be ensure all the cutter tightened are not otherwise cutter may cause damage and also it creates more noise and vibration. Motor has been mounted on the other side of the frame with help of bolt and nut. Hopper can be mounted on the cutter assembly to feeding of coconut leaves properly. Then mount the pulleys and v- belt to set the belt proper tensioning otherwise slipping of belt occurs. Finally, all the assembly work is done machine is ready work.

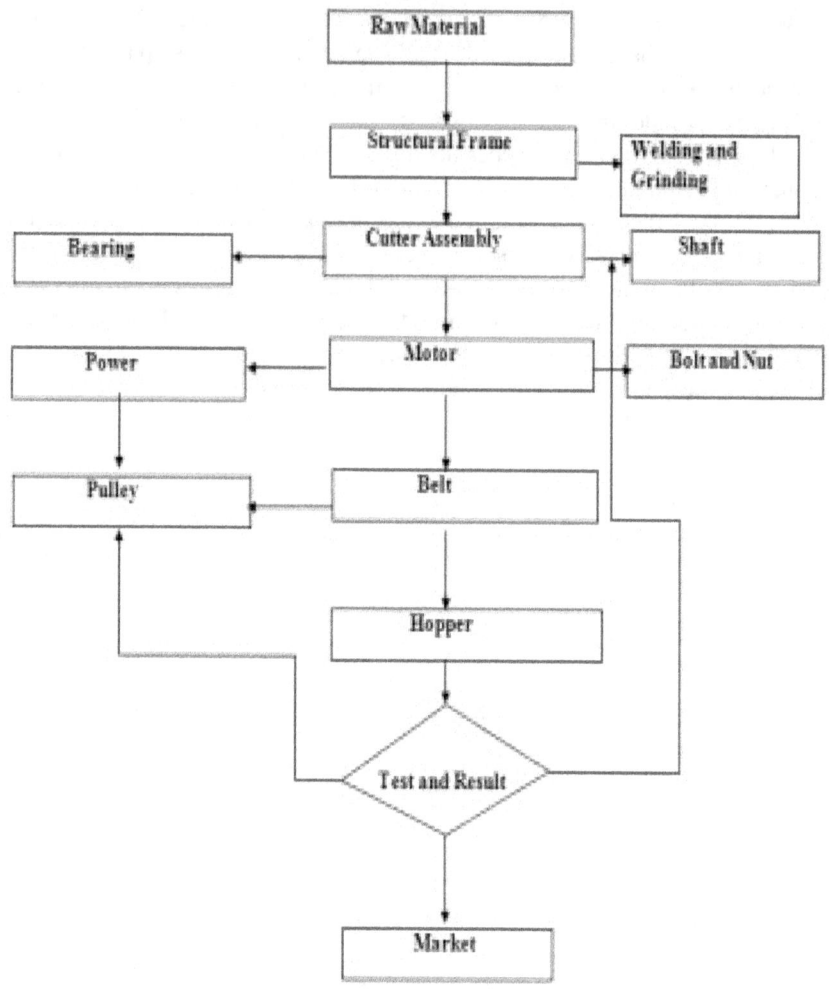

FIG 3.8 Flowchart

CHAPTER 4

FABRICATION PROCESSES

4.1 PROCESSES INVOLVED:

While fabricating the model we have incorporated the following processes:

1.CUTTING:

Cutting processes work by causing fracture of the material that is processed. Usually the portion that is fractured away is in small sized pieces called chips. The metal cutting involves the removal of excess material from the work-piece in the form of chip using a wedgeshaped performed this operation to obtain mild steel bars of suitable length for our frame.

FIG 4.1 GRINDER CUTTING MACHINE

2.DRILLING:

Drilling is a cutting process that uses a drill bit to cut a hole of circular cross section in solid materials. The drill bit is usually a rotary cutting tool often multi-point. We used this operation to fix at the bearings to the frame as it was not possible to weld them.

Fig 4.2 HAND DRILLING

3. WELDING

We have performed ARC welding for the fabrication purposes as it provides a Range of voltage and current which can beset according to the material being used.

ARC WELDING

Arc welding is a welding process that is used to join metal to metal by using electricity to create enough heat to melt metal, and the melted metals, when cool, result in a binding of the metals

5. BENDING

When sheet metallic is bent, it stretches in length. The bend deduction is the amount the sheet steel will stretch whilst bent as measured from the outside edges of the bend. The bend radius refers back to the inside radius. The fashioned bend radius depends upon the dies used, the material homes, and the fabric thickness

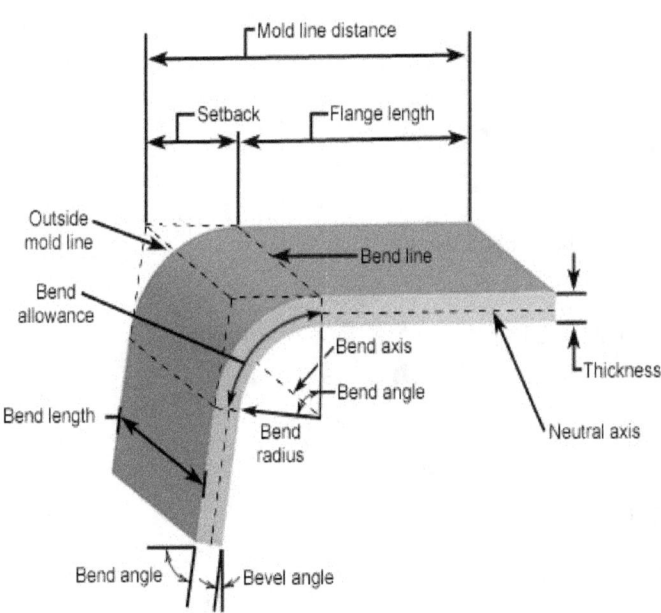

FIG 4.5 Bending Operation

4.2 Engineering Design standards

Engineering standards should be followed in each component of our system. In this section, each component that has been selected for the project is defined. The selected components are screws, gears, blades, and the motor. The screw standard has been taken according to ANSI metric. The gears standard has been taken according to ISO. The motor has been taken from Running-Motor Brand. Ronning-Motor is a company based in China specialized in manufacturing all kinds of universal motors, dc and induction motors.

4.2.1 Bolts and Nuts

Type of Screw and nuts have been used:

BOLT -HEXA HEAD BOLT:
- Length: 30 mm

NUT- HEXA NUT
- Diameter-5mm

This standard produce s that are easier to recycle because their relative size when compared to higher security standers of the micro-cut. DIN P-4 standard shredders produce approximately 400 particles. Furthermore, DIN P-4 provides shredded biodegradable waste.

4.2.3 MOTOR

FIG 4.9 Motor

The motor is manufactured by Running-Motor company, the specifications of the motor are

Its a fan motor with high rpm and torque.

Usage/Application For Fan Indoor & Outdoor

Phase   Single and Three Phases

Speed   1400-2800 Rpm

Power Source Type Electric

Voltage V   220-440 V

2.4.4 W1209 Digital Temperature Controller Thermostat Module

The W1209 is an incredibly low cost yet highly functional thermostat controller. With this module (HCTHER0006) you can intelligently control power to most types of electrical device based on the temperature sensed by the included high accuracy NTC temperature sensor.

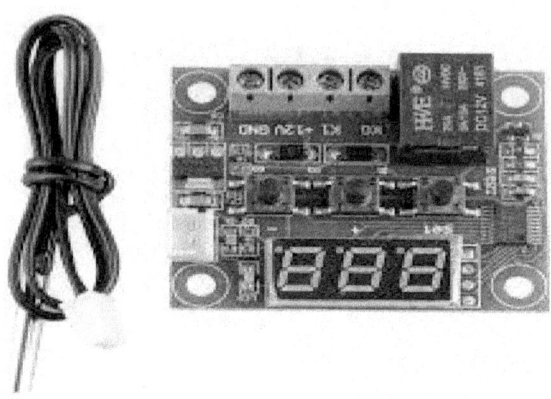

FIG 4.10 W1209 SENSOR

W1209 Specifications

Resolution at -9.9° to 99.9°: 0.1°C; 1°Cover 99.9° Measurement

Accuracy: 0.1°C; Control Accuracy: 0.1°C.

Refresh Rate: 0.5 Seconds.

Input Support Power (DC): 9V – 12V. ...

Measuring Input: NTC (10K 0.5%)

4.3 Theoretical Calculations.

Motor angular velocity = 2800 RPM

1 RPM to rad/sec = 0.10472 rad/sec

2800 RPM = 293.21 rad/sec

Operating voltage of motor = 220v

Current = 3 ampere

Power = 3*220 = 660 watt

Power will remain constant

Power = torque * angular velocity

Torque = power/angular velocity

Torque = 660/293.21

= 2.5 N-m

4.4 RESULTS

A compact portable Compositing machine was fabricated. This machine handled 1 kg of organic waste per day and converted into compost in 14 days. 14 kgs of organic waste was managed and near about 5 kg of organic compost was produced. There was volume reduction due to loss of moisture. Temperature sensor was set at

35 degrees of temperature rose above it the cooling fan cooled the composting bin and provided appropriate temperature for composting. Shredder provided a high surface area of material which was feed in it. All waste including vegetable, fruit waste and hard food waste was shredder properly. Motor operated well on 220 volts supply. Vibrations produced were less and machine operated properly. Motor rpm was noted using tachometer and it was approx. 2800 RPM. **Torque** produced by motor was 2.5N-m. There was some wearing in sprockets during shredding operations. Machine is easy to operate and safety was properly monitored to avoid any mishappening.

CHAPTER 5

PROJECT ANALYSIS

5.1 Life-Long Learning

In this project, all the team members have learned new things in the research process. The team has learned more details about gears, blades, and shafts. The team also learned about the types of paper shredders and different levels of security. Moreover, the team learned how to divide work, manage time, and overcome obstacles. The experience that the team acquired is very valuable and is sure to help in any and all future projects.

5.1.1 Hardware Skills

In this project, one of the hardware skills the team learned is calculating the rotational speed (RPM) of the gears using a Tachometer laser. The process is simply by using a special sticker that reflects the laser coming from the Tachometer while the sprockets rotate. The Tachometer is digital and thus the RPMs of the gears will show directly in the device. Moreover, the team has also used some electrical knowledge in order to install the cooling fan in the shredder.

5.1.2 Time Management Skills

Time management is significant in any project or plan. The team members tried to balance the project and its requirements, as well as the other course duties and requirements. In every meeting, the team members declared their status regarding their availability during a specific period. The team members carefully followed the Gannt chart to complete each milestone in a timely manner. The methods that were used to communicate and manage tasks were WhatsApp, Discord, and Google meetings. Moreover, the team communicates directly with the advisor once each task .

5.1.3 Project Management

The team members were committed to always be in touch in order to manage the project and its tasks. In each week, the team members gather in Google meetings or Discord to plan for the coming task. The team might have three or four meetings in a week depending on how urgent the case is or how heavy the task is, and some meetings lasted for over four hours. In the end, the team successfully finished all the tasks on time.

5.2 Impact of Engineering Solutions

This Project has multiple benefits in different aspects. It will be useful to society as well as the economy, and it will be extremely useful for the environment.

5.2.1 Society

This project is useful for society, especially in the household section, and is very. This project will help in terms of waste management. The houses and apartments can use it for composting.

Apartments can have a sperate area where this machine can be set up for waste management, and residents can take compost for gardening.

5.2.2 Economy

This project is helpful to the economy, as it helps in production of value added product from waste product by recycling biodegradable waste and encourages the continued safe disposal of it and its reuse to produce a new product in an environmentally safe way. When compost is produced in large amount it can sold to make profit.

5.2.3 Environment

This project plays a significant role in saving the environment. It greatly helps in protecting the environment, land, air, and sea by cleanly disposing of waste, which eliminates the need for disposing waste to landfills or using other harmful methods of disposal. So, shredding paper and disposing of it in the right way, and choosing to recycle, will help save the earth's environment and its inhabitants. Millions of liters of methane can be stopped from going into atmosphere. Methane is a potent global warming gasworks is suffering from climate change, so green and clean solution are needed. IPCC reports say making cannot go back but can make efforts to save earth.

5.3 Contemporary Issue

Thus, there is an essential need to dispose of biodegradable waste in the right manner, especially that recycling bins are not widely used in the country except in Aramco facilities that have special bins for recyclable materials. This project might be a good solution to dispose of biodegradable waste properly until the recycle /composting bins become everywhere.

CHAPTER 6

DISCUSSION AND CONCLUSION

6.1 Conclusion

In conclusion we both chose this project gladly with persistence to complete it and to learn from it, and to help people and society. Our team have used their knowledge from different courses, especially in Dynamics, Mechanical Engineering Design, and Manufacturing Methods. The team applied the skills of researching effectively, applied time management skills. Moreover, our team was able to put its special fingerprint on this project by installing the Temperature sensor along with fan and to fabricate the shredder on our own and build the prototype accordingly. When it comes to the difficulties, our team was able to find a suitable design for the shredder and was able to find a good workshop.

6.2 Future Recommendation

There are various ways to make this project better. The first one is to select materials with better mechanical properties than plywood. The project could also be better when using different gearing systems and increasing the AC motor power and decreasing the angular velocity. Choosing an outer body made of metal will be much better than using an outer body made of plastic. Finally, the team recommends using mechanical shredders when there is a need to dispose of household waste properly without damaging the environment or creatures

REFERENCES

[1] Anu. (n.d.). Helical gears or spur gears? Retrieved November 21, 2020, from https://clr.es/blog/en/spur-gears-helical-gears

[2] Compactor Management Company. (2020, July 27). Ways Paper Shredding Promotes Sustainability and Save the Environment. Retrieved October 15, 2020, from https://www.norcalcompactors.net/paper-shredding-sustainability-save-environment/

[3] Coombes, A. (n.d.). Shred on Site. Retrieved October 14, 2020, from https://shredonsite.co.uk/Blog/Paper-shredding-can-help-build-customer-trust-in-theseways

[4] History of Shredding Machines. (2020, September 28). Retrieved September 29, 2020, from https://accushred.net/blog/shredding-machine-history

[5] How Paper Shredding Promotes Recycling, Sustainability and Circular Economy. (2019, January 23). Retrieved October 14, 2020, from https://greenbusinessbureau.com/blog/how-paper-shredding-promotes-recyclingsustainabilityand-circular-economy/

[6] Mechanical Shredder - Forestry Biological Sciences ... (n.d.). Retrieved from https://www.ceias.nau.edu/capstone/projects/ME/2015/MechanicalShredder/index_files/Final Report.pdf

[7] Ogbeide, Nwabudike, & Igbinomwanhia. (2017, February 2). Design and Development of an Electric Paper Shredding Machine. Retrieved November 2, 2020, from https://www.google.com/url?sa=t&rct=j&q=&esrc=s&source=web&cd=&ved=2ahUKEwjny4GforTsAhWt4YUKHdUKBk0QFjAAegQIARAC&url=http%3A%2F%2Frjees.com%2Fdownload%2F%3Ffile%3DV02-N02-546-562.pdf&usg=AOvVaw3kq1L6DuDn31QKmA53vEYl [8] Oreko, B. U., Emagbetere, E., Oghenekowho, P. A., & Oghenevwaire, I. S. (2019, May 5). Design and Construction of a Paper Shredding Machine. Retrieved November 2, 2020, from http://www.jmest.org/wp-content/uploads/JMESTN42352953.pdf

[9] Paper Shredding Facts. (2013, February 08). Retrieved September 29, 2020, from https://www.shredinstead.com/papershreddingfacts#:~:text=The%20first%20paper%20shredder%20was,approved%20on%20Augu st%2031%2C1909.

[10] Says:, C., Says:, R., & Says:, R. (2020, August 27). Paper Shredder Security Levels DIN 66399. Retrieved November 21, 2020, from
https://www.recycling.com/paper-shreddersecurity-levels-din-66399/

[11] Slocum, A, 2008, "Power Transmission Elements II", Fundamentals of Design .http://pergatory.mit.edu/2.007/resources/FUNdaMENTALs%20Book%20pdf/FUNdaMENTALs%20Topic%206.PDF

[12] Ul Haq, E, Ibrahim, U, Moghees, A, 2020 "Paper Shredder Design - me3053." StuDocu www.studocu.com/en-us/document/capital-university/machine-design-and-cad/other/papershredder-design/8517780/view.

[13] Woestendiek, J. (2018, December 08). The Compleat History of SHREDDING. Retrieved September 29, 2020, from https://www.baltimoresun.com/news/bs-xpm-2002-02-10-0202110302-story.html

www.ingramcontent.com/pod-product-compliance
Lightning Source LLC
Chambersburg PA
CBHW070957220526
45471CB00007B/3072